BOHR & QUANTUM THEORY

Paul Strathern was born in London and studied philosophy at Trinity College, Dublin. He was a lecturer at Kingston University where he taught philosophy and mathematics. He is a Somerset Maugham prize-winning novelist. He is also the author of the *Philosophers in 90 Minutes* series. He wrote *Mendeleyev's Dream* which was shortlisted for the Aventis Science Book Prize, *Dr. Strangelove's Game: A History of Economic Genius*, *The Medici: Godfathers of the Renaissance*, *Napoleon in Egypt* and most recently, *The Artist, The Philosopher and The Warrior*, which details the convergence of three of Renaissance Italy's most brilliant minds: Leonardo Da Vinci, Niccolo Machiavelli and Cesare Borgia. He lives in London and has three grandchildren.

In THE BIG IDEA series:

BOHR & QUANTUM THEORY

The Big Idea

PAUL STRATHERN

arrow books

Reissued by Arrow Books 2010

1 3 5 7 9 10 8 6 4 2

Copyright © Paul Strathern, 1998

First published in Great Britain in 1998 by Arrow Books

The Random House Group Limited
20 Vauxhall Bridge Road, London, SW1V 2SA

www.rbooks.co.uk

Addresses for companies within The Random House Group
Limited can be found at:
www.randomhouse.co.uk/offices.htm

The Random House Group Limited Reg. No. 954009

A CIP catalogue record for this book
is available from the British Library

ISBN 9780099238324

The Random House Group Limited supports The Forest
Stewardship Council (FSC), the leading international forest
certification organisation. All our titles that are printed on
Greenpeace approved FSC certified paper carry the FSC
logo. Our paper procurement policy can be found at:
www.rbooks.co.uk/environment

Typeset by SX Composing DTP, Rayleigh, Essex

Printed and bound in Great Britain by
CPI Cox & Wyman, Reading, RG1 8EX

CONTENTS

INTRODUCTION

According to the great German theoretical physicist Werner Heisenberg: 'Bohr's influence on the physics and physicists of our century was stronger than that of anyone else, even Einstein.' And Heisenberg ought to have known – he spent a good deal of his life discussing (and arguing fiercely) with both of them.

Bohr's greatest achievement was to solve the riddle of atomic structure by applying quantum theory. This resulted in major scientific advances – and major scientific bafflement. How? No one still really knows what quantum theory is.

The great pianist Vladimir Horowitz once remarked of Mozart that he was 'too easy for beginners, too hard for experts'. The same applies to quantum physics, according to Bohr's colleague and biographer Abraham Pais. So if even the simplified version contained in this book begins to baffle you, at least you can console yourself that you're getting somewhere. Put very simply (for those of us who find even easy Mozart too difficult) quantum theory states that *particles at sub-atomic level do not obey the laws of classical physics. Indeed, such entities as electrons can exist as two different things simultaneously — matter or energy, depending upon how they are measured.*

The main trouble with quantum theory is that it is quite simply unbelievable. It has nothing to do with common sense. But 20th century science is far more exciting than common sense (which Einstein sensibly dismissed as 'the accumulation of prejudices we have acquired by the age of 18').

Bohr became director of the Institute for Theoretical Physics at Copenhagen, and from here he as good as masterminded the golden era of quantum physics. This took place in the 1920s, involving many of the finest young scientists of the post-Einstein generation. Together and separately, through discussion and individual brilliance, these pioneers opened up a field which a quarter of a century earlier no one had even suspected existed. The effects of this age of discovery have proved a mixed blessing. We now have a fair idea of how the world of physics works – from the most minute sub-nuclear particles to black holes. We also know how to destroy it with a nuclear holocaust. Bohr lived long enough to assist in the making of the first atomic bomb. When he realized what he had done, he spent the rest of his life campaigning against it.

LIFE & WORKS

Niels Bohr was born on 7th October 1885 in Copenhagen. He came from a distinguished Danish family. His father Christian Bohr was professor of physiology at Copenhagen University, and almost won the Nobel Prize for his pioneer work on the chemistry of the respiratory system. However, his most lasting effect on Danish society was due to his love of England and all things English. Christian Bohr was a great believer in football, and played a missionary role in the establishment of this popular religion in Denmark. Niels's mother Ellen was of Jewish descent, her family being prominent in banking and political circles. Despite his Christian christian name Bohr senior didn't believe in religion, and the family atmosphere was progressive-liberal-intellectual. From all accounts, the Bohrs were a pleasant, understanding, tolerant family. One after another their friends later recalled how pleasant, understanding and tolerant they were. It all sounds too good to be true – or utterly stifling. It's difficult to judge which.

Practically the only anecdote we have concerning Niels's childhood is characteristically unrevealing. One day Christian Bohr pointed out a tree to his young son, indicating how beautifully its trunk divided into branches, which then divided into thinner branches, which then produced leaves. The child Niels is said to have replied: 'Yes, but if it wasn't like that there wouldn't be any tree.' Psychologically and symbolically, as well as literally, this says practically nothing. Or alternatively, may be viewed as pregnant with all manner of significance. Take your pick: it's all we have.

Passing from blank childhood to bland youth Niels grew to be a quiet unassertive young man. Photos of this period depict a tall well-dressed but somehow rather lumpish adolescent, in a starched wing collar, with a decidedly hangdog expression. His chubby cheeks sag, he has large fat lips, and a faintly guilty look in his small close-set eyes. His speech was halting, and there was little in his manner which hinted at a first class brain. It was a different story at school. Here he was bright (but not brilliant), and quite willing to use his bulk when it came to fights. He soon became known for his dour strength, and shone at sports. Much to his father's pride, he was soon selected for the school football team. Following in his father's footsteps he took a keen interest in science. He was particularly drawn to experimental work in the labs, where he managed to combine exceptional skill with natural clumsiness. Smashing glass tubes and flasks was his forte. On one occasion when a series of explosions emanated from the labs, his long-suffering teacher was heard to remark: 'Oh, that must be Bohr.'

But by far the most important thing in Niels Bohr's life was his younger brother Harald, who joined him at the prestigous Gammelholm Gymnasium in Copenhagen. A fellow student remarked of their relationship: 'I have never known people to be as close as they are.' The brothers were inseparable. Harald was one and a half years younger than Niels, but quickly began to catch up with his bright and sportingly able brother. Niels showed no sign of resenting this. By the time they were leaving school, the younger Harald was already surpassing his brother on all fronts. He was not only a brilliant mathematician but also a highly accomplished footballer. As a stylish mid-field player Harald soon outshone his brother's occasional acrobatic feats in goal. From the start the younger Harald was the lively witty brother, whilst Niels had a quieter, more prosaic manner. Yet despite such rivalry and differences in temperament the brothers' deep rapport remained unruffled. Nothing ever came between them.

Or so we are led to believe. Bohr has become a national monument in Denmark, and the flesh and blood human being has been squashed beneath this solid Bohr edifice. It is difficult to believe that there wasn't at least some hidden tension in this unusually close brotherly relationship. Few older brothers give up their starring role without some resentment. Yet the brothers appear to have accommodated each other with remarkable sensitivity and psychological understanding. They divided the world of scientific learning between them. Harald appropriated mathematics, whilst Niels devoted himself to physics. This way they could consult each other, and even help each other, without any rivalry. Likewise, the goalkeeper's position on the team is unique, unchallenged by any other player (no matter how brilliant).

Even the Bohrs' fraternal relationship at home appears to have been devoid of the sulks and fisticuffs of normal brotherly affection. And this idyllic picture is completed by their older sister Jenny, who was equally brilliant. She went on to study at Copenhagen University and Oxford, before returning to Denmark to become an 'inspiring' teacher who was known for her 'warmth'. Yet despite this 'warmth' she never married, and we are sadly informed that 'nerves gave her difficulty in later life'. The truth is not so pleasant. The beloved sister and first child of the Bohr household was soon to become an incurable psychological wreck, who ended up locked away in a provincial lunatic asylum. According to her death certificate she died of 'manic depressive psychosis in its manic phase', a clinical formula which nonetheless evokes a chilling picture. As Harald was to admit at her funeral: 'from her earliest youth on, she was hampered, often made powerless, by illness [which] took all her strength.' The presence of such a frightening figure in the Bohr family home casts a somewhat different light on the unnaturally close relationship between Niels and Harald.

This insistence on a psychological dimension to Niels's relationship with his brother is not gratuitous. It is surely no accident that Niels's later work was characterized by its essential *ambiguity*. Quantum theory is about the compatability of two apparently irreconcilable opposites. And two of Niels's most important theoretical conceptions were Correspondence Theory and the Principle of Complementarity, both of which stress similarity despite underlying difference. Niels Bohr understood the notion of ambiguity at the most profound level, and his great scientific work sought to resolve this in harmony.

In 1903 both brothers entered Copenhagen University. These were exciting times. A new century was beginning, and the world was on the brink of changing beyond recognition. In the same year the first motorized taxi cabs appeared on the streets of Copenhagen. In America the Wright Brothers made their first flight, and Marie Curie received the Nobel Prize for the discovery of radioactivity.

Both Bohrs were soon playing on the university football team Akademisk Boldklub, one of the strongest teams in the country. (Later known as AB, it still flourishes in the second Danish league.) Niels sporadically excelled himself in goal – though when the play was down the other end he often passed the time making calculations in pencil on the goalpost. Indeed, it was his absorption in these mathematical pursuits, as much as the shooting skill of the opponents, which forced him to make some of his most spectacular saves. Harald, on the other hand, shone without such prompting, and eventually appeared for Denmark at the Olympic games (where they beat France 17–1 in the semifinal, then lost 2–0 to England).

By now the symbiotically close brothers were inspiring (or goading) each other to their full intellectual potential. Niels's painstaking temperament seemed ideally attuned to complexities of physics, and his accomplishments in his chosen field were soon matching those of Harald in mathematics. Even their fellow students began using the word 'genius' when referring to either of the Bohr brothers. Niels was always an avid reader, and did his best to keep up with the latest achievements in science. He quickly gained renown amongst his classmates for correcting some of the physics textbooks. What they said was quite simply wrong – and he was able to prove it, with evidence from recent discoveries.

During their student years the Bohrs continued to live at home. Here their father regularly invited some of the finest minds in Denmark to dinner. Copenhagen was no longer a provincial backwater: the previous generation of intellectuals had included the philosopher Søren Kierkegaard, pioneer of existentialism, and the critic George Brandes, the man who discovered Nietzsche. After dinner Bohr senior and his intellectual colleagues would take part in philosophical discussions – which Niels and Harald were allowed to attend as 'silent listeners'. (This may have been a progressive household, but it seems that even grown-up children were still expected to be seen and not heard.)

In 1907, during the final year of his degree course, Niels Bohr won the Gold Medal of the Royal Danish Academy of Sciences and Letters. This was for an essay on the surface tension of water. It was an astonishing feat for an under-graduate, and marks his first real emergence as a scientific super-intellect. Bohr completed the experimental work for this essay in between studying for his final exams, and almost missed the deadline set by the Academy. (The surviving manuscript shows that parts of it were hastily copied out by Harald, presumably from Niels's notes.) Niels's experimental work involved the precise analysis of the vibrations in a jet of water. Every one of these experiments was assembled and carried out by Bohr alone, and required a jet of water with a mean radius of under a millimetre. To produce this he needed long glass tubes of a similar proportion with an elliptical cross section, which he blew himself. The speed of the water in the jet was measured by cutting it twice at the same point at a given interval, then measuring the length of the cut segment by photographic means. The vibrations (the waves formed on the surface of the water) were also measured by photography. Most of his experiments had to be carried out during the early hours when the streets were empty, so as to avoid even the smallest disturbing vibrations from passing traffic.

This was to be the only original experimental work which Niels Bohr carried out entirely by himself. It bears all the hallmarks of his precise, painstaking approach and brilliant analysis — together with a near-miraculous absence of bungling. Even Bohr realized that this was a once-off. The strain of working with so much fragile glassware obviously took its toll, and from now on he always made use of physically adept collaborators when undertaking experimental work.

Being forced to listen, without being allowed to reply, has long been recognized as a stimulant to radical thought. Bohr's enforced silence at his father's after-dinner philosophical discussions was to prove no exception. He began thinking for himself. And in contrast to the learned discussions he endured in silence, Bohr's philosophical thinking proved astonishingly original from the start. Indeed, it even anticipates certain aspects of Wittgenstein's thinking. Bohr found himself disturbed by how a word can be used to describe a state of consciousness (eg, drunkenness), and at the same time be used to describe the external behaviour which accompanies this inner state. He saw that when a word referred to mental activity it was essentially ambiguous. (Here, for the first time, the concept of ambiguity appeared in Bohr's thought: tellingly, it is both profound and unresolvable.) In an attempt to resolve this ambiguity, Bohr drew an analogy with mathematics. He compared these ambiguous words (such as drunkenness, anger, joy and so forth) to multivalued functions. Put simplistically, these functions can have different values at the same point – but this ambiguity can be overcome by specifying to which 'plane' the value refers. Imagine a three–dimensional axis (see page 21):

If the intersecting point can have three different values, we can overcome this difficulty by assigning each value to a different plane: a, b or c. Bohr suggested that this method could

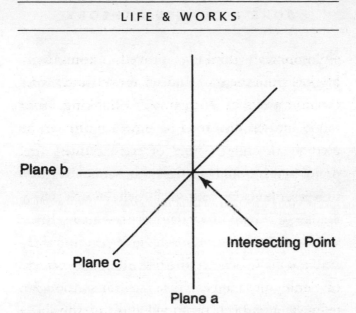

also be used in the philosophical problem of ambiguous words. When a word had conflicting meanings, we could overcome this by naming which 'plane of objectivity' it referred to. For example, drunkenness could refer to mental confusion or lack of physical coordination – two separate things. By specifying which 'plane of objectivity' the word referred to (ie, in this case, 'subjective' or 'objective') we could remove this ambiguity.

Alas, history is littered with great scientific (and philosophical) minds vainly trying to reduce language to a precise logical structure. But as we all know to our cost, conversation is seldom a logical procedure. (The great 17th century German philosopher-scientist Leibnitz devised a scheme which reduced all moral arguments to mathematics: at the end of the discussion the points were totted up, and the winner was the one with highest score. If only life was as simple as geniuses would have it . . .) Bohr's suggestion was equally ingenious, and equally doomed. He was forced to accept that such ambiguities are inherent in language. He was beginning to understand how conflicting interpretations can exist simultaneously. (This notion uncannily prefigures quantum theory, as can be seen by referring to the simplified definition in italics on page 8.)

Such forays into philosophy are usually disastrous for scientists. Philosophy concerns itself with how things are, science applies itself to how things work. In other words, science ignores philosophy and just gets on with it. But there are times when even science has to clean up its act and ask itself what on earth it's doing. The turn of the 20th century was one of these rare times. A previous one had occurred during the

17th century, when Galileo suggested that science had to conform to reality, not just ideas. The truth was arrived at by experiment, not just by thinking about it. This method had reached its apotheosis with the great scientific discoveries of the 19th century. But with the coming of the 20th century it somehow didn't seem to be enough. Inadequacies were being exposed in this empirical approach. The most glaring of these had occurred in 1905, when Einstein put forward his Special Theory of Relativity. Contrary to the old view of science, relativity *had* been discovered by just thinking about it. (Einstein used mathematics, not experiment.)

Something utterly new was in the air. Science was beginning to question itself, and its methods. What was it? What was it doing? Bohr's interest in philosophy was echoed by many leading scientific minds of the period. (Einstein read the 18th century Scottish philosopher Hume, who questioned the notion of cause and effect; others studied the 19th century German philosopher Kant, whose epistemology attempted to explain the nature of space and time.) Science was being put to the test of philosophy – and what emerged would provide the basis for the greatest scientific age of them all: the 20th century.

The spectacle of science turning on itself was exciting in the extreme. Everything was up for grabs. The Viennese philosopher–scientist Ernst Mach (after whom the sound barrier speed is named) even went so far as to question the existence of the atom. Almost half a century earlier the Russian Mendeleyev had revolutionized science with his Periodic Table of atomic elements – but was this all based on a false assumption? What exactly was an atom? Who had ever actually seen one? According to Mach the atom was no more than some outmoded concept, a hangover from previous unscientific thinking. Science had saddled itself with a meaningless abstract idea, which no one had ever actually observed. Nothing was sacred any more: even the most fundamental notions could be questioned.

And soon Bohr was doing just this. In 1909 he started on his doctoral thesis. This turned out to be a purely theoretical work, without any experimental input by its author, which came to an earth-shattering conclusion. (Bohr evidently decided that earth-shattering experimental work was best left in more adept hands.) The title of Bohr's thesis was 'An Investigation into the Electron Theory of Metals' (*Studier over metallernes elektrontheori*). Bohr rejected the extremity of Mach's position. Not only did the atom exist, but we were at last beginning to understand something about what it was. By the last years of the 19th century most scientists (but not Mach!) considered the atom to be the fundamental form of matter. Indeed, its very name comes from the Greek word atomos, meaning 'indivisible'. Then in 1897 the British physicist J.J. Thomson discovered the electron – the first known sub-atomic particle. The electron had a negative electric charge. It looked as if the atom had an internal structure. Thomson suggested that the atom was like a spherical cake, of positive electric charge, which was embedded with enough raisin-like electrons to neutralize this charge.

This picture was somewhat modified when it came to explaining things such as magnetism in metals. According to the electron theory of metals, a metal could be pictured as a gas of electrons amidst a lattice of positively charged ions.

In these early days, when no one really knew precisely what they were talking about, an ion was defined in circular Alice-in-Wonderland fashion as 'an atom which has lost its electrons'. Tentative theory was building upon tentative theory. Mach did indeed have a point. What was the foundation of all this theory, when still no one had ever actually seen an atom?

Regardless of such strictures, Bohr theorized forward into the unknown. In his thesis he argued with painstakingly precise analysis that magnetism revealed inadequacies in the electron theory of metals. The obvious answer was to question the electron theory – but Bohr took an entirely different tack. Here for the first time we glimpse the astonishing originality and daring of his thought. Bohr argued as follows. The electron theory of metals brilliantly accounted for almost all the qualities exhibited by metals – the only trouble was it started going wonky when it tried to account for the *quantities* involved in metallic behaviour. For instance, when some metals were placed in an electric field the strength of their magnetism (which was dependent on electrons) did not accord with the laws of classical physics. But what on earth did this mean? Bohr put forward an amazing conjecture. It seemed that at sub-atomic level the old assumptions of classical physics just didn't work. In order to describe what went on within the atom, it looked as if an entirely different kind of physics might be called for.

But this was impossible. It was as if Bohr was suggesting that at sub-atomic level waterfalls flowed upwards and fairies existed. Here indeed was an Alice-in-Wonderland world where 2 + 2 didn't necessarily equal 4. Even Bohr wasn't sure what all this meant.

So he decided to head for the source. After Bohr finished his thesis in 1911 he set off for the Cavendish Laboratory in Cambridge to study with J.J. Thomson. If anyone knew about the behaviour of electrons, surely their discoverer would. But in Cambridge Bohr found himself running up against a brick wall. This was partly of his own making – but not entirely so. Thomson just wasn't interested in this overbearing enthusiastic Dane, who couldn't even express *himself* properly in English, let alone his complicated theories. Bohr hardly helped his case with his faulty use of terminology. For instance, when talking about electricity he would refer to 'load' when he really meant 'charge'. (Bohr was fluent in Danish and German, but apart from this he was always an enthusiastic linguist, rather than a skilled one. In later life he insisted upon greeting the French ambassador with a cordial: 'Aujourd'hui!') The low point of Bohr's Cambridge experience came during an informal discussion with Thomson and his colleagues. As Bohr was trying his best to explain a point concerning the electron theory of metals Thomson interrupted him, dismissing his ideas as rubbish – and then proceeded to say precisely the same thing himself using different words. Matters weren't improved when Bohr later insisted that some of the great man's calculations concerning the electron were faulty. 'They have all lost confidence in me,' he wrote home forlornly. But he was possessed of the pig-headedness of genius. It definitely looked as if there was another form of physics, which contradicted the laws of physics. He had no idea where his illogical ideas were leading, but he was determined to persist with them.

Bohr realized that he had to learn English properly, so with characteristic perseverance he sat down with the works of Charles Dickens and a large Engelsk-Dansk Ordbog. Looking up every word he didn't understand, he began making his way through descriptions of even stranger behaviour than that of electrons – the foibles of Victorian England as described in *The Pickwick Papers, Oliver Twist, and Martin Chuzzlewit.*

Then came Bohr's lucky break. In October 1911 he attended the annual Cavendish dinner, held in honour of the 19th century founder of the Cavendish Laboratory (a distant relative of the great 18th century chemist Henry Cavendish who first calculated the weight of our planet). Over the port, the Cavendish dinner was addressed by Ernest Rutherford, whose work on radioactivity had already led to revolutionary speculations about the nature of atoms and their structure. Bohr was entranced by this bluff middle-aged New Zealander, whom one of his colleagues described as 'a charming blend of boy, man and genius'. (It is perhaps no accident that Bohr's father had died prematurely a few months previously.) Rutherford was equally struck with Bohr. He liked active men, and was impressed when he learned that Bohr was a footballer of professional standard. (In true Kiwi style, Rutherford had been a robust rugby player for his college at Cambridge.) More importantly, Rutherford understood what Bohr had to say – despite his English, which remained halting and torturous when expressing complex scientific ideas beyond the range of Mr Pickwick. ('Engelsklish', as a friend called it.) In Rutherford's opinion: 'This young Dane is the most intelligent chap I have ever met.' Some accolade, coming from the leading physicist of the age – Bohr's worth was at last beginning to be recognized. Rutherford invited him to come and work on his research team at Manchester University. Bohr accepted with enthusiasm, writing home: 'I shall have wonderful conditions in Manchester.'

Bohr and Rutherford may have established an immediate rapport, but they were in fact opposites. Rutherford was the supreme practical physicist, and despite his robust approach was capable of devising experiments of extreme subtlety. He demanded downright proof, and despised lofty theorcticians who wouldn't stoop to sullying their hands with laboratory equipment. Bohr, on the other hand, was reared more in the French and German tradition. He preferred subjecting hypotheses to rigid logical analysis in order to discover their implications, and was never afraid to build theory upon theory. At this stage Continental reason prevailed over British empiricism in Bohr's approach. And for once, Rutherford didn't seem to mind. 'Bohr's different. He's a football player,' he told his colleagues.

Rutherford's researches into radioactivity had led him to some bizarre conclusions. He had discovered that atoms were capable of disintegrating. Atoms of one element could fall apart to become atoms of a completely different element. Such conclusions were simply laughed out of court by classical physicists of the old school. This was nothing more or less than a regression to medieval alchemy, with its belief that base metals could be turned into gold.

Rutherford decided to investigate the internal

structure of the atom, and devised a series of experiments with his German assistant Hans Geiger (inventor of the Geiger counter, used for measuring radioactivity). Rutherford basically agreed with Thomson's picture of atomic structure: a uniformly positive spherical cake embedded with negative electron raisins. But his experimental work on the radioactive decay of atoms had led him to visualize the positive cake as more like a sphere of positive gas studded with tiny electrons.

Rutherford realized that the only way to

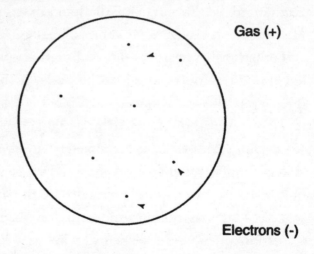

Rutherford's initial model of atomic structure

investigate something as miniscule as an atom was to bombard it with something even smaller, namely a sub-atomic particle. Fortunately, he knew that when radioactive atoms disintegrated into smaller atoms they emitted a stream of sub-atomic particles. He chose to use one type of these, called alpha particles. They were much larger than electrons, and had a positive charge.

Rutherford set up an experiment in which a beam of alpha particles from a radioactive source was fired at a thin sheet of gold foil. When the alpha particles had passed through the gold foil they struck a fluorescent screen, producing tiny flashes.

Rutherford was not surprised when most of the alpha particles passed straight through the

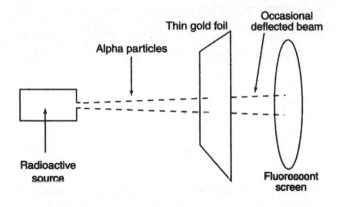

sheet of gold foil, as he assumed the atoms were mostly gas. Occasional alpha particles were deflected, and this was obviously due to them striking one of the tiny electrons in the atomic sphere of gas. The electrons were so small that they only deflected the large speeding alpha particles by around 1°. But as the experiments continued, he was surprised to discover that some of the alpha particles were being deflected by as much as 10°. Yet how could this be – when the bombarding alpha particles were comparatively huge, were travelling at vast speed, and the electrons were so small?

A few days later an even more astonishing event occurred. It was found that a few of the alpha particles were actually bouncing *back* from the sheet of gold foil. 'It was quite the most incredible event that ever happened to me in my life,' commented Rutherford. 'It was almost as incredible as if you fired a fifteen-inch shell at a piece of tissue paper and it came back and hit you.'

What on earth was happening? What could possibly stop, and bounce back, a sub-atomic particle of such momentum? It didn't take long for a mind such as Rutherford's to come up with an answer. The atom obviously consisted of a tiny but extremely concentrated centre, which carried a positive charge. This would thus withstand the momentum of the positive alpha article, and also repel it (like two positive ends of magnets placed together). Rutherford was now able to come up with a model of the atom. According to his picture, the atom was almost entirely empty. At its centre it consisted of a miniscule but extremely dense nucleus, which occupied a billionth of its space. (Imagine a pea in a cathedral.) This positive nucleus was surrounded by a number of negative electrons, which circled it in fixed orbits, under the influence of its attraction.

The nuclear atom was born (to be followed by nuclear physics, and the nuclear age). This was one of the most aesthetically pleasing ideas ever conceived in science. It looked as if the smallest unit in the world worked in the same way as the solar system. The micro and macro worlds mirrored one another!

Rutherford's conception was inspired. It had just one drawback — it didn't work. According to the laws of classical physics the orbiting electrons would simply radiate away all their energy in a millisecond and collapse into the nucleus. Also, what stopped the positive particles which made up the nucleus from repelling one another (again, like the magnets)? A central nucleus consisting of positive particles would simply fly apart. These questions were unanswerable. Rutherford's atomic structure was unstable.

But Bohr thought otherwise. Never one to be put off by the laws of science, Bohr set about trying to find a solution which supported Rutherford's 'impossible' atomic structure. Bohr was convinced that the solar system picture of the atom was too good not to be true. It explained so much that had previously been inexplicable – and Bohr was one of the first to see this.

Each of the elements in the Periodic Table had an atomic number, which indicated its position in the table and thus its properties. This number (closely related to its weight) obviously reflected the number of positive units of electrical charge in the nucleus. These would in general be equalled by the number of negative electrons orbiting the nucleus. The physical and chemical properties of any atom were evidently dependent upon the orbits of these electrons. The ease with which it could either take on or shed electrons would depend upon the position of the orbit. Different types of orbit might even explain how there were different classes of elements, and how their properties recurred as they did in the Periodic Table. Why, Rutherford's model could explain *everything*. It *had* to be right – even if it meant denying the laws of physics! (After all, Bohr had found himself questioning these before, in his thesis on the electron theory of metals.)

To explain how Bohr managed to solve this apparently insurmountable problem, it is first necessary to understand precisely what the laws of physics were. At this time (around the turn of the 20th century) the laws of physics were based on the fundamental assumptions of classical physics. This was the scientific view of the world developed largely by Galileo, Newton and later Maxwell. It saw the universe as a vast and complex machine, operating on strictly mechanical lines, in absolute time and absolute space. All motion within this universe had a cause. The sequence of cause and effect was rigid and immutable. Everything was determined (and was thus amenable to explanation). There were two forms of energy – that possessed by particles in motion, and that which travelled in waves (such as light, radiowaves etc). The former behaved like colliding billiard balls, the latter like waves travelling on the surface of the sea. These two forms of energy were mutually exclusive. As explained by Newton's theory of gravity, bodies were attracted to each other according to their mass and distance. And as explained by Maxwell's electromagnetic wave theory, light travelled in waves, with different frequencies producing different types of radiation (eg, different colours). These basic laws and assumptions of classical physics were utterly certain and beyond doubt. This was simply the way the world worked.

And who could deny this? Since the foundations of classical physics had been laid by Galileo in the 17th century they had instigated an age of scientific advance unparalleled in previous history. From a world where witches were burnt at the stake and the Earth was the centre of the universe, to the early skyscrapers of Chicago, the

Wright brothers' first flight, gravity, electricity, the combustion engine, the speed of light – classical physics explained them all.

Or so it appeared. But by the end of the 19th century certain cracks had begun to appear in this edifice of certainty. These were generally thought to be anomalies – caused by the inceasingly rapid, but uneven advance of scientific knowledge. Such hiccups would soon be ironed out by future discoveries.

Then in 1900 the German Max Planck, professor of physics at Berlin University, made an astonishing discovery. As he explained to his son one cold December morning while they walked in the woods near Berlin: 'Today I have made a discovery as important as that of Newton . . . I have taken the first step beyond classical physics.' But what exactly had he discovered?

One of the anomalies which had puzzled the classical physicists was the so-called 'ultra-violet catastrophe'. This came about as follows. A black body absorbs all frequencies of light, so when heated it should radiate all frequencies of light. But it doesn't: it emits frequencies at the lower range, which only gradually increase with the intensity of the heat. As we all know, a heated body emits low-frequency red light. Next, as it gets hotter, it emits higher frequency orange. Then yellow. Yet according to the laws of classical physics it should emit an equal amount of *all* frequencies of light. However, if this *did* happen we would all suffer from severe high-frequency ultra-violet burns every time we sat in front of a fire. The 'ultra-violet catastrophe' simply didn't happen!

Planck had managed to work out what *did* happen. But to do so he had been forced to contradict the laws of classical physics. He decided that electromagnetic radiation (light) behaved both as waves *and* particles. These wave-particles he named 'quanta' (from the Latin meaning 'quantity').

Low frequency radiation

High frequency radiation

According to Planck the low-frequency electromagnetic radiation (light) consisted of small quanta, and the higher frequency radiation consisted of larger quanta. When a body was heated, it took comparatively little energy (heat) to form the smaller quanta, which made up the low frequency red light. The larger quanta needed a greater amount of energy to form, and thus were only radiated as the heat increased. And even then, only a few to begin with.

To recap: why must quanta be waves *and* particles? Light has a frequency, therefore it must travel in waves. Yet how to explain the absence of the 'ultra-violet catastrophe'? Planck suggested that the waves were not continuous (as they should have been, according to classical physics). Instead they were quanta, ie, bundles of waves or objects, which were emitted like discrete particles. And the quanta for different frequencies of light each had a different physical size (quality of matter), which explained why ultra-violet was not initially emitted so intensely. Quanta were thus waves that travelled in particles, and particles that consisted of waves. Impossible, according to classical physics, where waves are continuous, and particles consist of matter, not waves. (For waves are merely oscillations at a certain frequency which pass *through* a medium or substance – like waves passing *through* water.)

Planck worked out that the size of the quanta varied in proportion to the frequency of the radiation. He expressed this in his famous formula:

$$E = hv$$

where E is the energy value of the quantum and v is the frequency of the radiation. The h is a fundamental constant, now known as Planck's constant. (A fundamental constant is a physical quantity, which can be expressed as a number, and it is always the same no matter the circumstances, anywhere in the universe. Another example of a fundamental constant is the speed of light. Planck's constant has since been calculated to have a value of 6.626176×10^{-34} joules. A miniscule sum, almost zero in fact. But only the fact that it is *more than zero* means that high frequency quanta require more energy, which thus rescues us from the 'ultra-violet catastrophe'.)

Planck's idea was so revolutionary that at first no one could accept it. Even he found it difficult to believe. But five years later Einstein confirmed Planck's 'quantum theory', as it became known, by using it to explain the photo-electric effect, another anomaly in classical physics. This effect took place when ultra-violet light struck certain metals, causing an emission of electrons. These electrons didn't behave according to the laws of classical physics. The rate of their emission depended upon the frequency of the bombarding light, rather than its intensity. The higher the frequency, the more electrons were dislodged. This could be explained if the light was regarded as quanta (with higher-frequency light consisting of larger quanta). Once again, it seemed light travelled as both waves and particles.

However, most of the scientific world refused to accept such nonsense. This quantum theory was just an ingenious 'German invention', which bridged a gap where our knowledge had temporarily oustripped any comprehensive theory to explain it. Quantum theory was simply illogical – it would never last. Within a few years a more comprehensive theory would properly account for the anomalies explained in such complex fashion by quantum theory. And this comprehensive theory would doubtless accord with classical physics. Even Planck and Einstein themselves were convinced that this was what would happen.

Einstein's explanation of ultra-violet light striking certain metals was in much the same scientific territory as Bohr's thesis on the electron theory of metals. Bohr too had come to the conclusion that sub-atomic particles didn't obey the laws of classical physics. But neither he, nor anyone else, could see how all this linked up with the problem of atomic structure. And that was the problem Bohr now faced in defending Rutherford's 'solar system' model.

A key to the structure of different elements was discovered as a result of a new type of spectroscope invented in 1814. This had been the work of the Bavarian optician Joseph von Fraunhofer, who had survived being buried alive when the tenement building in which he was working collapsed on top of him. All others in the building had been killed, and the Elector Maximillian Joseph marked his miraculous survival by awarding him the princely sum of 18 ducats. This eventually enabled von Fraunhofer to undertake the independent research which led to the invention of the spectroscope. If any element in gaseous form is intensely heated it glows, and when this emitted light is examined with a spectroscope it is broken down into its component colours. These show up as bands of coloured lines in the overall spectrum of colours, and it was found that each element had its own characteristic 'emission spectrum'.

A typical emission spectrum

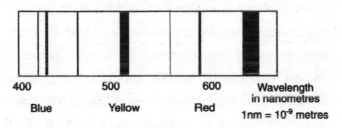

| 400 | 500 | 600 | Wavelength in nanometres |
| Blue | Yellow | Red | 1nm = 10⁻⁹ metres |

Each line in the emission spectrum could be assigned a precise number-value in accordance with its wavelength. These appeared to repeat, much like a harmonic sequence.

But a formula for this sequence continued to elude all comers until 1885, when one was discovered by the Swiss Johann Balmer, a teacher in a girls' secondary school in Basel. Balmer worked out the following formula: 'Square the number 3. Divide 1 by the result and subtract this fraction from ¼. Multiply the answer by the number 32,903,640,000,000,000,000.' This gave you the frequency (and thus the wavelength) of the red line in the spectrum for hydrogen.

Just the sort of things you'd expect from a provincial schoolteacher (which was perhaps why it had eluded more brilliant minds). But this formula wasn't quite so clumsy as it appeared. If instead of opening it with the number 3 you began with the number 4, you obtained the green line in the spectrum, and with 5 you obtained the frequency of the violet line. And when more lines in the hydrogen emission spectrum were later discovered, the next numbers (ie, 6, 7, 8 and so on) were found to give figures which also uncannily matched their frequency. Balmer's formula worked, but no one knew why. Even more intriguingly, no one had any idea what these line spectra meant. It was just accepted that each element had its own 'signature'.

Amazingly Niels Bohr was almost completely ignorant of spectroscopy when he began working on Rutherford's solar system model of the atom. According to the legend, one day Bohr was glancing through a schoolboy physics textbook and happened to come across Balmer's formula. Puzzled by its intractable uniqueness, he began playing with the figures in his head. Then came the 'Eureka!' moment. He recognized that Balmer's formula could be written in another way, using the constant *h* which Planck had used in his formula for establishing the size of the quanta in his theory of light. *This meant that the spectrum of an element was linked to quantum theory!*

But how, precisely? Bohr concentrated on the simplest element of all – hydrogen. According to Rutherford's picture of the atom, hydrogen had just one electron orbiting its central nucleus. By the laws of classical physics this should have radiated just one band of colour. Instead it radiated several separate ones, connected in a fixed regular pattern (according to the 3,4,5 and so on in Balmer's formula).

The single electron in the hydrogen atom could theoretically circle the nucleus in any number of different orbits. The greater the radius, the faster the electron orbited, and thus the greater its energy. Smaller orbits would have lesser energy. So if the electron moved to a smaller orbit, it would have to emit energy. This was emitted in the form of light. Now the energy emitted by an atom made up its spectrum. Bohr saw at once what this meant. The precise energy bands which made up the spectrum were somehow linked to the different orbits of the electron. Bohr realized that in order to emit the fixed energy required, the electron would have to move between certain fixed orbits. When the electron had a fixed orbit, the atom would be in what Bohr called its 'stationary state'. Only when it moved between these fixed orbits would it emit the precise energy required to produce a particular band in the spectrum. Because of this, he understood that the fixed orbits of these stationary states would have to be linked to Balmer's formula (because this gave the figures for the bands in the emission spectrum).

As we have seen, in an inspired moment Bohr had realized that Balmer's formula could be rewritten with Planck's constant, which concerned the behaviour of quanta. *This meant that the different orbits of the electron in the hydrogen atom were fixed according to quantum theory!* The structure of the atom was determined by quantum theory.

To recap: in each stationary state the electron

had its different fixed radius. Bohr had already realized that these states must be related to Balmer's formula for the spectral bands. So each fixed electron radius was related to the different applications of Balmer's formula!

In other words, Bohr had realized that Balmer's formula for spectral bands *and* quantum theory were connected to the structure of the atom.

Now Balmer's formula for placing the spectral bands begins: 'Square the number 3. . .' (or 4 for green, 5 for violet, or even 1 or 2, and so on). This meant that the fixed electron radii were related in precisely the same way. So it was now comparatively easy for Bohr to calculate the radii of these orbits.

The diagram on the next page shows Bohr's model of the first four orbits of the electron in the stationary states of the hydrogen atom.

The radius of each orbit is related to the number required for Balmer's formula. The difference between the radii gave the difference between the energy states, which in turn matched the energy bands on the spectrum.

The difference between these orbits may have produced the correct figures for the hydrogen

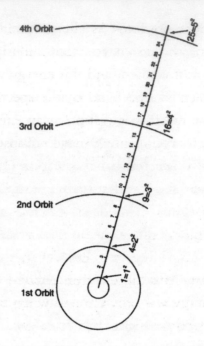

spectrum, but how was the energy actually released? Easy, according to Bohr. The electron simply 'jumped' from one orbit to another.

Imagine a train travelling on a fixed track (the electron on its fixed orbit). Contrary to all the laws of physics the train suddenly jumps from one track to the inner track beside it. In doing so it releases a flash of light (a band of colour is released for the hydrogen spectrum).

According to the laws of classical physics this

was just impossible. As already mentioned, the electron should have continuously radiated energy as it orbited, and this energy would only have been in one band of the spectrum. Also, with the electron radiating away energy in this fashion it would simply have collapsed into the nucleus in less than a nano-second (10^{-9} sec). So obviously classical physics just couldn't account for what was happening. On the other hand Bohr's model did – even if it contradicted all the rules. According to Bohr's model, in the stationary state the electron emitted no energy. The energy was only emitted when the electron jumped from a higher orbit to a lower one. (And energy would be absorbed when it jumped from a lower to a higher orbit.) So not only the structure of the atom, but also the behaviour of its subatomic particles, appeared to conform to quantum theory.

Bohr's picture was a curious mixture. It obeyed many of the laws of classical physics (eg certain elements of the dynamics of orbiting bodies), whilst at the same time contradicting others (such as the laws of causality). However, it had one sensational feature. *It agreed precisely with Planck's original quantum theory of light.* The light emitted by an electron jumping from one orbit to another would not be continuous. Instead it would be emitted in bursts, precisely like Planck's 'quanta'.

The diagram below shows Bohr's picture of the hydrogen atom, with the different radii of the electron. The arrows show the various jumps between orbits which produce the different

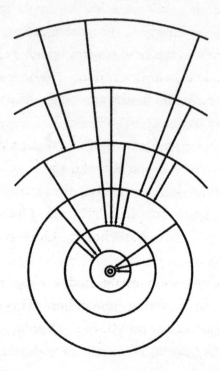

frequencies in the spectrum of hydrogen.

Quantum theory could explain how the spectrum of the hydrogen atom was emitted. This, together with Einstein's explanation of the photo-electric effect, seemed to confirm Planck's original notion of quantum theory. But more than that, it considerably extended Planck's idea. Bohr showed how quantum theory was essential to the understanding of sub-atomic phenomena. And in doing this he came up with the first comprehensive picture of a stable atomic structure. Here was the new physics which explained the anomalous behaviour of sub-atomic particles.

Bohr began his work on atomic structure in 1912 after moving from Cambridge to Manchester University to work with Rutherford. (By one of the whimsies of fate, another major 20th century thinker was taking the train in the opposite direction. In 1912 the philosopher Wittgenstein left Manchester University for Cambridge.)

Rutherford soon established a close rapport with his new young researcher. Though he wasn't too keen on Bohr's chosen line of research: the attempt to come up with some solid backing for his (ie, Rutherford's) unstable solar system picture of the atom. Rutherford considered his picture to be 'provisional'. It was too theoretical and unproven to be relied upon; in his view you couldn't build upon such a notion. But building on theory was Bohr's speciality – and he seemed to have an infallible nose for a theory which would last. Despite Rutherford's reservations, he was soon encouraging Bohr with the same bluff enthusiasm (and keenly perceptive questioning) to which he subjected all his team. In the mornings Rutherford was in the habit of marching through the laboratories singing 'Onward Christian Soldiers'. These were exciting times in science – and for those involved it felt very much like a crusade. The gap between the new young world and the monolith of time-honoured certainties was widening throughout European culture. This was the age when Europe ruled the world, the last great days of the Austro-Hungarian Empire – but it was also the age of cubism, relativity and silent movies.

At the end of the summer term Bohr left Manchester for Denmark. Two years previously his beloved brother Harald had left to take up a post at Göttingen University, whose reputation for mathematics was second to none. It was during this period of unaccustomed loneliness that Niels met a fluffy-haired blonde student called Margarethe Norlund. Niels and Margarethe were entranced with each other from the start. She was soon helping prepare the final copy of his PhD thesis, and they became engaged before Niels returned to Manchester. They appear to have succeeded in creating an utterly banal relationship, as dull (to outsiders) as it was profound. The concept of 'true love' has long since become extinct in progressive societies – rendered nauseous by saccharine sentimentality, exploded by psychiatry, sabotaged by divorce. But in the early years of the 20th century true love was still alive and cooing, and expected to last for a lifetime.

When Niels finally returned home to Copenhagen in the summer of 1912, he married Margarethe. Photos of the couple show a surprisingly youthful looking Niels, his heavy sensual features lightened by a smile. Beside him leans his transparently happy young wife, her arm comfortably linked in his. They look utterly at ease with each other – a rapport that was to last for 50 years.

After his marriage Bohr was appointed a junior lecturer at Copenhagen University, but maintained close ties with Manchester. He corresponded regularly with Rutherford and his colleagues about his progress. Bohr's final paper on atomic structure underwent several revisions. Some of these were in answer to technical points raised by his correspondents in Manchester, but many were due to Bohr's method of com-position. He was a painfully slow writer, always searching for a more precise meaning. He liked to immerse himself in words, rather than simply use them. It was as if his creative process took place in language, rather than used language to express itself. (Despite structuralism, there is a distinction here – as between syntax and metaphor.) However, all this made for difficult reading. And when Bohr's technique was extended to the lecture hall it proved a severe handicap – to both speaker and listener alike. Fortunately this defect was counterbalanced by Bohr's evident enthusiasm and the sheer brilliance of his ideas. Understanding a lecture by Bohr became something of an intellectual feat, testing mental athletes to the full. Only those possessed of marathon concentration and sprint comprehension could last the pace.

In March 1913 Rutherford received the final version of Bohr's paper. After reading it, Rutherford wrote back: 'I suppose you have no objection to my using my judgement to cut out any matter I may consider unnecessary in your paper.' Bohr caught the first boat across the North Sea and was in Rutherford's office next morning before he could finish the first verse of 'Onward Christian Soldiers'. All day and long into the night Bohr defended his paper line by line. Eventually the middle-aged Rutherford buckled under the persistent insistence of the earnest young Dane. The paper was eventually published uncut in the *Philosophical Magazine*, where it caused a sensation. Overnight Bohr became the *enfant terrible* of the atom.

Bohr had suggested that quantum theory was the way atoms worked. His ideas were dismissed as nonsense. How could the fundamental basis of matter possibly rest on something so utterly unstable? For many, quantum was still just that too clever 'German invention', which would never last. Surprisingly, they even ridiculed it in Göttingen. Gamely, Harald Bohr stuck up for his brother: 'If Niels says something is true, it must be true.' But the leading German theoretical physicist Max von Laue was so incensed by Bohr's ideas that he announced: 'If this theory is correct I shall quit physics.' (Fortunately he was dissuaded from such drastic action, and later even became a close friend of Bohr's.)

This, and similar apoplectic reactions, were not entirely a result of pig-headedness or prejudice. Many felt that science simply couldn't go on like this, without destroying itself. Bohr's explanation was quite simply *unscientific*. It not only defied the laws of classical physics, but it was illogical too. To describe the structure of the atom with a combination of classical physics and quantum theory was absurd. The principles of classical physics and the principles of quantum theory were *contradictory*.

Worse still, for Bohr, it soon became clear that his model for the hydrogen atom was not complex enough to account for some of the finer details in the hydrogen spectrum. This problem was eventually to be solved by the professor of physics at Munich, Arnold Sommerfeld, the first significant believer in quantum theory after Einstein. Sommerfeld applied Einstein's relativity theory to the movements of the electrons about the central nucleus, and realized that these orbits must be elliptical. (In fact, more like the real solar system.) This eventually led to the following picture of the nucleus and the different fixed orbits of the electron.

Sommerfeld's model accounted for all the

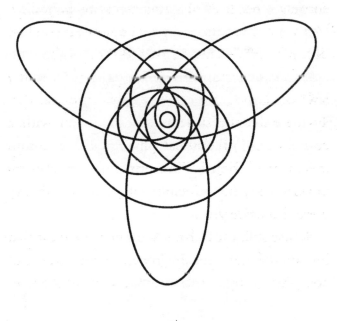

extra fine lines in the hydrogen spectrum. But more serious problems remained. The fundamental contradiction between quantum theory and classical physics was proving intractable. Also, Bohr's picture may have fitted the hydrogen atom, which had just one electron – but difficulties soon arose when it was applied to more complex atomic structures. Was this quantum hydrogen atom just a freak one-off? Bohr struggled with the horrendously complex theoretical work involved in trying to extend his plan of one particular atomic structure to the general scheme of *all* atomic structures.

This work required much time and energy on Bohr's behalf. When in 1914 Rutherford offered him a post back at Manchester, with no tiresome lectures involved, Bohr readily accepted. But before he could take up his post the First World War had broken out. Denmark remained neutral, but there were naval skirmishes off the Danish coast between British and German warships. The British soon assumed control of the North Sea and assured safe passage to neutral shipping. Even so, the ferry on which Bohr and his wife set sail for Britain was forced on a long detour, through storm and fog, around the coast of Scotland.

Bohr had always believed in the internationalism of science: this was knowledge for the benefit of mankind, not for separate nations. His point of view was echoed by several of the rising generation of German scientists. Despite the barriers of war, news from Germany soon began reaching Bohr in Manchester. He learned that already theoretical predictions made from his quantum atomic structure were producing spectacular advances in the new science of sub-atomic physics. Although the unanswered questions about Bohr's structure remained, it was becoming apparent that it explained too much to be abandoned. Bohr was excited by these developments. So much so that he now took the unusual step of trying to confirm one of the German predictions by means of a practical experiment.

The complex apparatus needed for this experiment required intricate glassware, and Rutherford made sure that Bohr obtained the services of the best man available. This turned out to be a German called Otto Baumbach. Only

Rutherford's immense prestige could have ensured that an enemy alien had the free run of his laboratory during a war. Unfortunately Otto the expert glass-blower soon became ill-at-ease working alone amongst the enemy. In times of stress he developed the habit of bawling a tirade of anti-British propaganda, informing the astonished scientists precisely what would happen to them when the Germans won the war and took over the laboratory. Unlike the others, Bohr remained unconcerned: 'All good glass-blowers are temperamental,' he remarked. 'He is merely relieving his feelings with this super-patriotic nonsense.'

In the end the situation became too much for all concerned, and after a series of inflammatory incidents Otto was finally led away to be interned. Unfortunately this was followed by an even more inflammatory incident when Bohr's apparatus caught fire, reducing Otto's irre-placable glassware to a heap of exploded splinters. Once again, Bohr's efforts in the laboratory were doomed.

As the huge importance of Bohr's work was gradually realized, the Danish authorities began to comprehend their loss. In an unprecedented move the University of Copenhagen offered the 30-year-old Bohr a professorship. There was also a promise that if he returned, funds would be made available for him to set up a special institute where he could continue with his research.

In 1916, as the British and German fleets were preparing to confront each other off the coast of Jutland, the Bohrs made the hazardous journey back across the North Sea. Margarethe was now pregnant, and a few months later the Bohr's first child was duly born in Copenhagen. The atheist Bohr named his son Christian, after his atheist father. 'This will always be a remembrance of your stay in Manchester,' wrote Rutherford, when he received the news. But Bohr's remembrance of Manchester was in fact to be even more profound. Rutherford had filled the gap in Bohr's life left by the death of his father. The somewhat naive and earnest young Bohr had soon come to regard the hearty kindly great man as both a father figure and a role model. Bohr would never forget Rutherford's Manchester laboratory. This was how science should be done: in an atmosphere of camaraderie and fruitful argument, in which all took part. If young Bohr had spoken, the great Rutherford had listened. When Bohr came to set up his own promised institute, he would not forget this. And the effect on science was to be incalculable. It is no exaggeration to say that Bohr's encouragement of such an atmosphere in his own laboratories would revolutionize 20th century science.

But Bohr's informality was not appreciated by all who met him. On becoming professor of physics at Copenhagen University, Bohr was expected to present himself in morning suit and white gloves for an audience with the king. Bohr duly turned up on the appointed day and was formally introduced to King Christian X, a somewhat crusty military man who was a great stickler for court etiquette. The king cordially shook Bohr's hand, saying how pleased he was to meet such a great footballer. Bohr felt obliged to point out that *he* wasn't the Bohr who was the famous footballer – that was his brother. The king was dumbfounded. One of the cardinal rules of court etiquette laid down that you *never* contradicted the king, no matter what he said.

Christian X decided to overlook this grave breach of decorum, and give Bohr another chance. So he simply began all over again, saying how pleased he was to meet such a great footballer. Bohr decided to try a different tack. He was used to explaining things to the mighty: he always managed to make Rutherford understand in the end. Yes, he was a footballer, he agreed – but it was his brother who was the *famous* footballer. Outraged, the king pronounced: '*Audiensen er jorbi!*' (The audience is finished.) And Bohr was at once motioned to take his leave, in the usual fashion – backing away from the royal presence towards the door.

Despite this gaffe, the Danish authorities kept to their promise. In 1918 work began on the new *Institut for Teoretisk Fysik*, with Bohr appointed as its future director. Despite its name, the Institute was also to house a number of expensively equipped laboratories, where practical work could be carried out. Initial funds were provided by Carlsberg, brewers of probably the best Danish beer. (In Denmark charity begins in the brewery: Carlsberg supports science, Tuborg the arts.) But as we all know, beer alone cannot provide complete support. Bohr himself was forced to hunt up funds for his new institute, before it finally opened in 1921.

The Institute quickly began attracting ambitious young scientists from all over Europe, each keen to learn from and work with Bohr. As a result, exciting advances were soon being made in Copenhagen. On the basis of Bohr's atomic model, it had been possible to predict the existence of a new hitherto unknown element – element No. 72. By means of spectral analysis, element No. 72 was finally identified for the first time at the Institute in Copenhagen. The discovery of this new element did much to confirm Bohr's work, and it was duly named hafnium (after the Latin version of København).

However, this discovery quickly ran into difficulty. No sooner had it been made than it was disputed. A counter-claim was made by the celebrated 76-year-old Irish experimentalist Arthur Scott, who announced that he had already discovered this element nine years previously in 1913. He had even named it celtium, after his homeland. The dispute was quickly taken up by the popular press, becoming a matter of national pride in both countries. Hardly had Bohr and his Institute gained an international reputation, than it looked as if it might be ruined by a charge of cheating. Scott appeared in public brandishing a test tube containing a sample of celtium. Things reached such a pitch that Rutherford was called in to adjudicate. Rutherford eventually prevailed upon Scott to send a sample of celtium for spectral analysis in Copenhagen. Alas, this was found to contain no trace of element No. 72. Some elements have a half-life, others an even briefer existence. Celtium vanished into the mists of chemical legend, and has not been seen since (even in Ireland).

Later in 1922 Bohr was to receive the ultimate accolade – the Nobel Prize for Physics. These were the vintage years of the Nobel Physics prize. In the seven years from 1918, it was won by Planck, Einstein, Bohr, Millikan and Hertz. But not all were giants. In 1919 it was won by Johannes Stark, who later tried to purify Nazi Germany of 'Jewish Science' ie, relativity, quantum theory, nuclear physics and other such racially inferior rubbish.

Despite this public recognition, quantum theory was still very much in the melting pot. It was in the unique position of making great advances, yet still unable to build a certain foundation for itself. These were exciting times. Sometimes one of Bohr's brilliant young colleagues would ask him what direction he thought science was going. Bohr got into the habit of quoting Goethe's *Faust*: 'What is the path? There is no path. On into the unknown.'

Take, for instance, the frequency of lines emitted in the spectrum of an atom. According to the quantum picture, these were produced when the electron 'jumped' from one orbit to another, and no radiation was emitted when the electron orbited the nucleus in its 'stationary state'. However, classical mechanics said precisely the opposite. The radiation was produced by the electron as it orbited the nucleus, and could be calculated accordingly. Obviously both of these two pictures couldn't be true. Yet Bohr found that every quantum 'jump' picture could also be 'coordinated' with a corresponding classical mechanical orbit. The atom could be classical, or it could be quantum. Indeed, for lower frequencies quantum theory and classical mechanics came up with *precisely the same answer.* This led Bohr to formulate his famous correspondence principle. This stated that at sufficiendy low frequencies the laws of quantum theory and the laws of classical mechanics *become identical.*

This was of course nonsense. It was illogical, impossible, unthinkable etc. But it was what happened – and it was what worked! Little wonder that a scientist of the stature of von Laue had threatened to give up physics altogether if quantum theory proved to be right! Even Einstein clashed with Bohr. It was quantum physics which provoked Einstein to his famous remark: 'God does not play dice with the universe.' Bohr and his colleagues at the Institute in Copenhagen were in reality just as baffled. But they remained optimistic – they were determined to press forward along this increasingly wayward path. ('There is no path.') Even Bohr was aware that his correspondence principle was really just patching up the bits.

Yet these 'bits' were soon producing unprecedented advances in scientific knowledge. By the first years of the 1920s quantum theory was attracting the finest minds in science. Everything was up for grabs. Brilliant theories were even being produced by physicists who hadn't been born at the time Planck first proposed quantum theory. This was truly a 20th century science. And the major centre for it was Bohr's newly created Institute. (Indeed, for a while only Göttingen and Cambridge, where Rutherford had now moved, remained anywhere else in the race.) The list of co-workers at Bohr's Institute reads like an international roll call of the next generation of 20th century giants. The Swiss Pauli, Heisenberg from Germany, Dirac from England, Landau from Russia, and other less known geniuses – all at one stage worked at Bohr's Institute. And quantum theory now evolved into quantum mechanics, as the inner mechanics of the quantum picture of the atom were worked out. How did Bohr's atom actually *work*?

Quantum theory was flying by the seat of its pants. In the Bohr tradition, theory was building upon theory with bewildering speed. A piecemeal picture of immense complexity was being constructed of this sub–atomic world where neither logic nor causality held sway. But there were certain advances which transformed the entire picture. One of these was produced by the podgy 23-year-old Swiss prodigy Wolfgang Pauli, who was prone to deep melancholia whenever he discovered a problem he couldn't understand. One of these was an anomalous effect in atomic emission spectra which couldn't be explained by Bohr's model of atomic structure.

This model (sometimes known as the Bohr–Sommerfeld model) had now been considerably developed. It could already show the general arrangement of electrons about the nucleus in more complex atoms (ie, ones with more than the single electron in hydrogen). The electrons were arranged in the different fixed orbits about the nucleus. It was also found that these fixed orbits fell into separate groups, each of which was called a 'shell' (or layer). For instance, it was found that in every atom the inner shell consisted of just one orbit (represented in the following diagram by the innermost circular orbit). The next shell contained four orbits (as respresented in the diagram by one circular, and three elliptical orbits).

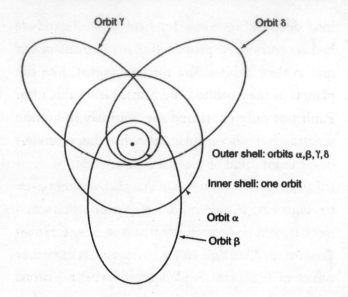

But according to Bohr's picture, the inner orbits should have become increasingly packed with electrons. (For energy to be emitted, an electron had to 'jump' to an inner orbit.) But for some reason the inner orbits didn't become packed with electrons. This problem was to be solved in 1924 by Pauli.

By now Pauli had become highly upset by his inability to explain the anomaly in atomic spectra which didn't fit Bohr's model – so much so that he had taken to wandering aimlessly through the streets of Copenhagen for hours on end, plunged into deep depression. But help was at hand. It had recently been proposed that electrons might *spin* as they orbited the nucleus (again, like the planets as they orbited the Sun). Using this idea, Pauli not only explained the anomaly in atomic spectra, but also explained why the electrons don't cram into the atom's inner orbits.

Put in simplified terms, Pauli showed that each fixed orbit could hold no more than two electrons. When this was filled, the next electron was *excluded* and had to pass into one of the other orbits in the shell. And if there were no orbits with a free space in this shell, it was further excluded and had to occupy an empty orbit in the next outer shell. This Pauli called the exclusion principle. Not only did it show how the electrons avoided all becoming crammed into the inner orbits, it also explained the structure of the elements in the Periodic Table.

For instance, the hydrogen atom had just one electron in the inner shell. The next element, helium, had two – thus filling the shell. This meant it was in a way more 'complete': it had little propensity to take on a new electron or lose one. It was thus unlikely to react with another element. This explained why helium had always been one of the so-called Inert Elements.

The third element, lithium, had two electrons in the inner shell, and one in the next shell. As we have seen from the preceding diagram, the second shell has four orbits, and thus room for up to eight electrons. When these are filled, we get an atom with ten electrons in all (two inner shell, eight outer shell). This, the tenth element, is called neon. Like the previously mentioned helium, it has completely filled shells. And both have similar properties: both are members of the Inert Elements. The periodically repeating similarity between the elements in Mendeleyev's original Periodic Table was now further explained. The properties of each element were dictated not only by the number of electrons it contained, but also by their disposition in the shells of orbits, and how much the outer orbit was filled.

Although Bohr did not collaborate with Pauli in devising the exclusion principle, he did play a part in its conception – as he did in so many of the advances during this golden era, which saw the birth of quantum mechanics (roughly

1924–28). All the time that Pauli was attempting to formulate his principle, he maintained a constant 'argument' with Bohr about it – by means of letters, and discussions during his stays at the Insitute in Copenhagen. In this way, Bohr became a kind of father-figure to the new advances in quantum mechanics. He didn't always agree with his younger colleagues. But the atmosphere of fruitful discussion between equals which he created at the Institute played a fundamental role in these revolutionary discoveries.

Another sensational development came in 1925, from the 23-year-old German whizz kid Werner Heisenberg, who besides being a brilliant physicist also found time to be an ace mountaineer, excellent pianist, could memorize inhumanly long tracts of Goethe, and seldom slept for more than a few hours. Despite such obstacles, he also managed to remain a human being – and soon established a close rapport with Bohr. It was Heisenberg who found a way of overcoming the illogicality of Bohr's correspondence principle. Both the quantum picture of atomic structure (the Bohr-Sommerfeld model) and the picture according to classical mechanics (Rutherford's solar system) were agreed that an atom emitted energy (light), which produced the atom's emission spectrum. But according to quantum theory this light travelled in quanta (energy bundles, or wave-particles). In other words, discontinuously. Quantum theory even explained why light travelled in this fashion: because each time the electrons 'jumped' from one fixed orbit to another, they released 'quanta' of energy. Yet according to classical mechanics, light was transmitted in continuous waves. That is, constantly. Now according to Bohr's correspondence principle, classical mechanics and quantum theory converged at low frequencies – to the point where they gave the same answers and were *identical*. But this just couldn't possibly be so. Either something is continuous, or it is intermittent – it cannot be both at the same time! (If my glass is continuously empty, it means there is never any beer in it. If it is intermittently empty, it means someone is filling it and I am emptying it.)

Heisenberg found a brilliant way of side-stepping this illogicality. Such problems as continuous/intermittent energy radiation were overcome simply by concentrating upon observation – *and observation alone*. Only the measurable properties of an atom were to be considered as 'real'. The concept of an atom as a minute solar system – whether it agreed with the Bohr-Sommerfeld or the classical model – was simply abandoned. As Heisenberg remarked: 'Why speak of an invisible electron orbiting inside an invisible atom. If they cannot be seen, they are not meaningful.' It didn't matter whether you pictured something as a continuous wave, or a discontinuous particle. This was irrelevant, if all you were concerned with was *measurement*. All measurements depended upon how they were taken, but the results couldn't disagree with one another. They were simply results.

This was a brilliant insight, but how were such *measurements* to be expressed in a meaningful form without a 'picture' to hang them on (ie, a model, such as the 'solar system' atom)? This was overcome by arranging the different measurements as rows and columns of numbers in matrix form. Then by applying matrix theory it would be possible to predict further values for physical variables (such as had applied to particles) and mathematical probabilities for varying energy states (such as had applied to waves). These rectangular rows and columns of figures proved much more useful than a 'picture' of an atom. They gave the first consistent form of quantum mechanics, allowing it to predict, in a manner resembling classical mechanics.

Now what was to happen? Was all the immense ingenuity of the solar system atomic structure simply to be thrown out of the window? But how could Pauli's exclusion principle, which helped account for so much in the behaviour of all the elements, simply be wrong? The answer was, of course, that it was not. In overcoming one ambiguity, quantum mechanics had merely thrown up another.

Two years later, Heisenberg came up with his famous uncertainty principle – which put the final nail in the coffin of classical physics. Following on from his idea that only measurement could produce certainty, Heisenberg realized that where quantum mechanics was concerned even this could not produce *overall* certainty. For instance, electrons were so minute that no matter how you tried to measure their behaviour, the way you measured it affected their behaviour. If you beamed light onto an electron, so you could 'see' it, this inevitably knocked the electron off course, affecting its velocity or its position. As Bohr put it: 'Any observation regarding the behaviour of the electron in the atom will be accompanied by a change in the state of the atom.' Yet without anything to measure the electron, we can't 'see' it at all – in other words we can know nothing about it!

In an experiment it was possible to measure the momentum of a sub-atomic particle, but you couldn't at the same time measure its position. And though you could set up an experiment to measure its position, you couldn't at the same time measure its momentum. This was the basic uncertainty of Heisenberg's uncertainty principle.

But it wasn't only the brilliant younger generation of European scientists who were responsible for the astonishing advances in quantum mechanics which occurred in the mid-1920s. Bohr was still producing creative work of his own. In 1927 he produced his 'principle of complementarity', which accounted for the position/momentum duality indicated by Heisenberg. According to Bohr's theory: 'evidence obtained under different experimental conditions cannot be comprehended within a single picture, but must be regarded as complementary in the sense that only the totality of the phenomena exhausts the possible information about the object.' In other words, there may be no single overall picture, but we can use a set of complementary ones which cover much the same ground. These are not logically contradictory because their realms of validity don't overlap.

This also effectively dealt with the particle/ wave duality. According to Bohr's principle of complementarity, whether an entity behaved as a wave or a particle simply depended upon what apparatus you chose for measuring its behaviour.

But serious questions still remained unanswered. Even after all possible information had been gathered, from all possible experiments, this was still not sufficient to make precise predictions. At quantum level you could never know *exactly* what was going to happen next – as you could in classical physics. It was possible to measure with extreme accuracy the momentum of a sub-atomic particle. It was also possible to measure its position with similar accuracy. But these couldn't be done *at the same time*. Yet only if they were done simultaneously was it possible to make an exact prediction about what it would do next. At quantum level such accuracy was impossible. Quantum mechanics worked on probability. You could only predict the likelihood of something happening.

Bohr countered these objections as follows. Agreed, it was impossible at quantum level to make precise predictions, as can be done for large-scale phenomena. But quantum mechanics was still capable of making surprisingly accurate predictions by means of probability. By taking large samples of electrons you could predict the general course of their future behaviour. It may have been impossible to say where one single electron would be in the future. But it was possible to predict, with considerable accuracy, the pattern behaviour of a large group of electrons. Such probabilities were simply statistical, rather than individual. Predictions at quantum level had to rely upon statistics and probability. These may not have been as simple and precise as classical mechanics, but for the most part they were just as effective.

This applied to all kinds of sub-atomic phenomena, from spectral lines to radioactivity. For example, in radioactivity the atomic nucleus disintegrates and emits an alpha particle. Quantum theory remained unable to predict precisely *when* a single nucleus would disintegrate and emit an alpha particle. However, by means of probability it was possible to show that a group of radioactive atoms would emit alpha particles *at a certain rate*.

In 1929 Bohr began a series of annual conferences at the Institute. These were attended by the top young scientists of Europe – many of whom had already spent working spells at the Institute. Here the difficulties of quantum mechanics were thrashed out. These were difficulties of immense complexity, at the cutting edge of science. And at the end of each conference, agreement was far from universal.

However, the moving spirit of these conferences would do his best to patch things up.

The earnest, pipe-smoking Bohr had a way of grinding down his opponent with the sheer persistence of his arguments, when he felt sure that he was right. This didn't work with the absent-minded Pauli, who would simply retire into his own mind and continue with his own thoughts, no matter how long Bohr bored on. With others it was different. Heisenberg, who was no wimp and could be equally sure that *he* was right, was on one occasion reduced to tears of frustration. (Afterwards, they patched things up by agreeing that the uncertainty principle and the principle of complementarity were merely the same thing expressed in different ways.) When things became too much even for Bohr, he would leap to his feet and rush out into the garden. The others would watch as he lit up a large cigar, and then set about vigorously 'weeding' the meticulously kept Institute flower beds – which sometimes reduced to a barren wasteland by the end of a conference.

The arguments were fierce and often deeply entrenched, but were conducted in a spirit of international cooperation. All present felt that they were attempting to get at the truth. This was what science was about, and this was what scientists did. The conferences at Bohr's Institute were far from unique. This was the era of great international scientific conferences, when a score or so of top minds would gather to discuss a particular scientific issue. Perhaps the best known of these were the celebrated Solvay Conferences, held at the Metropole Hotel in Brussels. These were financed by the Belgian industrial chemist Ernest Solvay, who had made a fortune out of the amonia-soda process (used in glass and soap manufacturing). At these gatherings, Bohr would be forced to defend his position against the likes of Einstein (who had actually received his Nobel prize for his work on quantum theory), Madame Curie (who had pioneered research into atomic structure with her discovery of radioactivity), and the combative Austrian physicist Erwin Schrodinger. Einstein and Curie were still convinced that quantum was essentially a passing phenomenon, a theoretical attempt to encompass anomalies which would one day one be resolved. Not so Schrödinger – who seemed intent on destroying quantum, but from within.

Schrödinger was lucky to be admitted to these

conferences at all. His eccentric habit of travelling dressed as a mountaineer, complete with heavy boots and rucksack, meant that he sometimes had difficulty getting past the door at the smart Metropole Hotel. But Schrödinger was capable of taking on more than doormen. When Heisenberg had produced his matrix mechanics in 1925, Schrödinger had immediately rubbished this and produced instead his own alternative 'wave mechanics'. (This set the cat amongst the pigeons, just as Schrödinger had intended, until it was later shown to be mathematically equivalent to Heisenberg's matrix mechanics.) More difficult to overcome was Schrödinger's attack on quantum's probability approach to prediction.

All this was odious nonsense, he declared in characteristic fashion. And proceeded to blow a nasty hole in probability prediction. This was achieved with his infamous example known as 'Schrödinger's Cat'. Schrödinger imagined a cat sealed in a box (see diagram on page 86). Inside the box there was also a radioactive source, a hammer, and a sealed glass flask of poisonous gas. When an atomic nucleus of the radioactive source disintegrated it released an alpha particle,

which released the hammer, which broke the flask, which released the poisonous fumes, which killed the cat.

Now quantum theory couldn't say what would happen to any particular radioactive atom, it could only predict the overall *rate* at which the atoms would decay. That is, the rate at which the atoms of the radioactive source would emit alpha particles. Say it predicts a 50% probability of one alpha particle being released per minute. Treating the box according to quantum theory, this means that at the end of a minute the cat is neither completely alive nor completely dead. Which is, of course, nonsense.

Schrodinger's cat also attacked the central notion of quantum theory – that a phenomenon had no value until it was measured (which of course depended upon *how* it was measured, which could even determine what it was viz a wave or a particle). Looked at from this point of view: after a minute in the box Schrodinger's cat remained in an intermittent state, neither dead nor alive – *until someone opened the box and looked inside.* Only *then* (ie, when it was measured) did it take on a value (ie, life or death). This patently wasn't true. Either the cat was already alive, or already dead. Measurement had nothing to do with it.

Schrödinger was convinced that his beast from the living dead would be too much of a horror story for quantum prediction, and its 'no measurement: no value approach'. But it wasn't. Despite the absurdity of Schrödinger's cat, quantum mechanics still relies upon predictability and insists that a phenomenon can have no value until it is measured. Likewise, Bohr's principle of complementarity (another object of Schrödinger's scorn) still remains very much the standard position of quantum mechanics.

However, further derivations of this principle have not stood the test of time. Bohr's principle of complementarity rather went to his head, and he began to see himself as the grand old man of knowledge (not just science). He rightly saw that the principle of complementarity was applicable to more than just quantum mechanics. In biology, for instance, there are two approaches. Biological phenomena can be classified from a functional point of view (for example: homo sapiens is a colonial zooid). But they can also be studied in terms of physical and chemical analysis (for example: vitamin C is essential to maintain life in the human organism). Instead of seeing these two approaches as opposed, Bohr argued that we would be better considering them as complementary. He then extended this approach to the thinner ice of sociology. Here he suggested that we should regard the study of human behaviour, and the (apparently opposed) analysis of hereditary transmission, as complementary when determining the main elements of a culture. Like almost anything to do with this so-called science, Bohr's suggestion was open to question. (Bohr's venture into this field also led him into dangerous racial speculations — but it should be stressed that these bore no resemblance to the dangerous theories which were coming to the fore across the border in Germany during this period.)

It wasn't until Bohr extended his principle into philosophy that he really made a fool of himself. Bohr's principle of complementarity raises a serious question for epistemology (the theory of knowledge, or what philosophical grounds we have for saying we know anything). But to claim, as Bohr does, that the ancient philosophers were in reality groping their way towards an expression of his principle, is plain daft. Quantum physics is a genuine problem for philosophy. Indeed, any philosophy which doesn't take account of it (and many still don't) belongs with the works of the ancient philosophers – fascinating antiquities, but inadequate for modern day experience. However, quantum physics (and with it the principle of complementarity) cannot *be* philosophy. It is science – which is human knowledge. Philosophy (in this case epistemology) studies the *grounds* of human knowledge.

Philosophy may have almost nothing to say nowadays (while science has everything to say); but it will always in this sense be 'deeper' than science. Even when conducted by the likes of Bohr. For once this brilliant conceptual thinker (one of the 20th century's finest) was out of his depth. As one of his colleagues commented (privately), Bohr was as suited to philosophy as he was to the open road: 'It was an act of faith to sit in an automobile driven by Bohr.' When he felt hot, he simply let go of the wheel and took off his jacket as if he was sitting in a stationary chair. Many lives were saved by the fact that his wife always insisted upon sitting beside him, and regularly made lightning adjustments to the wheel. (Despite the strength of the philosophical argument against Bohr, more recent advances in quantum physics have tended to support his case. Scientists now argue, with some justification, that quantum physics has simply hijacked epistemology, stripping philosophy – leaving it like the emperor in his new clothes. Yet, the unavoidable point remains: science and philosophy are not the same thing.)

Fortunately, Bohr only set up as a part-time guru. He continued at the same time to make major contributions to science. In common with many of his colleagues, he now turned his attention to the nucleus of the atom. In the late 1930s he put forward a model describing the make-up of the atomic nucleus, and how it worked. His theoretical picture helped clear up many of the conflicting ideas which were now emerging as a result of rapidly advancing experimental techniques. Here again, he viewed nuclear behaviour in terms of quantum mechanics.

Bohr saw the nucleus as consisting of a group of particles held together by short-range forces, much like the molecules in a droplet of liquid. When a particle strikes this nucleus/droplet, its energy can be quickly absorbed among the colliding particles and it becomes part of the droplet, which heats up accordingly. This state will continue for a long time (in nuclear terms), with the contained energy continuing to fluctuate randomly. The nucleus will decay only when the increased fluctuating energy causes a concentration of energy on a particle, allowing it to escape – much like evaporation taking place from a heated droplet. However, in the case of a large heavy nucleus (such as that of uranium), the escape of the particle will cause the droplet to split into two droplets of similar size. (This process is called nuclear fission.)

Bohr was the first to describe what actually took place during nuclear fission. He did this in

1939, just after it had first been discovered. This was as a result of an experimental programme carried out by German physicist Otto Hahn and his German-Jewish colleague Lise Meitner, who was forced to flee Nazi Germany in the midst of their work. Hahn secretly sent out the results of the completed experimental work to Meitner in Sweden, who analyzed them and realized that nuclear fission had taken place.

This news was passed on to Bohr, who realised at once the awesome implications of splitting the atom, and the vast amount of energy that would be released. On a visit to the United States later in the year he warned Einstein that Nazi Germany had the theoretical know-how to begin research into making an atomic bomb. As a result Einstein wrote to President Roosevelt, who quickly set up the Manhattan Project with the aim of making an American atomic bomb first.

In 1940 Denmark was overrun by Nazi Germany, but Bohr did his best to maintain the integrity of his Institute. The era of open international scientific cooperation was long since past, however scientific cooperation of a covert nature continued. Bohr remained in secret contact with British scientists, and in 1941 he was visited by his former colleague Heisenberg, whose relationship with Bohr had cooled somewhat as a result of historical developments. Heisenberg was one of the few top scientists to have remained in Nazi Germany. His work on 'Jewish Science' (ie, theoretical physics) meant he was vilified in Nazi scientific circles and referred to as a 'White Jew'. Nonetheless he had been offered (and accepted) a leading role in the German atomic bomb programme. At his meeting with Bohr in Nazi-occupied Copenhagen, Heisenberg passed on a cryptic diagram revealing how far the Nazi atomic programme had progressed. (The diagram was not of the bomb itself, but of a prototype reactor that didn't quite work.) Why Heisenberg took this action remains a matter of heated debate. Heisenberg later claimed that he wished to influence scientists on both sides to abandon work on the bomb. Bohr thought otherwise, and was deeply upset by this meeting (again, opinions differ as to precisely *why* he was upset). Bohr and his erstwhile favourite pupil never really got over what took place at this meeting.

By September 1943 Bohr heard that on account

of his open contempt for the Nazis he was due to be arrested. A hurried escape plan was organized. Bohr and his family travelled to a house in the suburbs of Copenhagen. At nightfall they crawled across a field to a deserted beach. Here they were met by a fishing boat, which ferried them 15 miles across the sound past the German patrol boats to neutral Sweden.

On arrival, Bohr was rushed to Stockholm. As soon as his escape was discovered, the German secret service was alerted with orders to assasinate him on sight. Although Stockholm was crawling with German agents, and there were many German sympathizers in official positions, Bohr managed to obtain an audience with the king. Prior to his flight he had heard that the transportation of Danish Jews to the camps was about to begin. The King of Sweden was persuaded to jeopardize Sweden's neutrality, and came out publicly against this grotesque violation of human rights. (Partly as a result of this, though more as a result of Danish ingenuity and courage, only 500 of Denmark's 8000 Jews were eventually transported.)

Several days later the British government

despatched an unmarked mosquito bomber to Sweden to pick up Bohr. By now German agents knew about both the bomber and Bohr's (separate) location. They were determined to do all they could to keep them apart. After a series of adventures and mishaps worthy of James Bond, Bohr was finally stowed away secretly in the empty bomb bay of the mosquito, which took off under cover of darkness. It managed to elude the Luftwaffe on its flight across Nazi-occupied Norway and headed out across the North Sea. By this stage the 57-year-old Bohr was in danger of freezing to death. When the mosquito at last landed safely in England, Bohr was found to be almost unconscious from hypothermia and lack of oxygen.

Bohr had scarcely been in England long enough to warm up (a process that could take some time in fuelless wartime Britain), before he was shipped to the United States. Here he immediately joined the Manhattan Project, travelling to the secret laboratories at Los Alamos where the bomb was being assembled. He was greeted by the American physicist Robert Oppenheimer, the leader of the project, and was able to show him and his colleagues the diagram which Heisenberg had passed on to him. But this didn't prove quite the reassurance Bohr had intended. To the Americans it was obvious that such a reactor couldn't work. Yet they felt that the Germans would immediately have realized this. Either that, or the diagram Heisenberg had passed on was an elaborate hoax, intended to disguise the real advances they had made. (This remains something of a mystery. The fact is the German effort to make an atomic bomb never really got off the ground. How much this was due to the deliberate efforts of Heisenberg is still a matter of fierce discussion.)

The first two US atomic bombs were dropped on the Japanese cities of Hiroshima and Nagasaki in August 1945, thus putting a full stop to the Second World War. Having played his part in creating the atomic bomb, Bohr was duly horrified when he saw its effect. He immediately began using his prestige to lobby at the highest level for all nuclear bomb research to be halted forthwith. His efforts were at best rebuffed. (Churchill was all for having him locked up.)

In 1945 Bohr returned to Copenhagen and his

beloved Institute. The Nazis hadn't really known what to do with the world's leading Institute, which specialized in racially unacceptable science. As a result, Bohr found things much as he had left them – including a bottle of auric acid solution which he particularly prized. According to the story, when Bohr had fled Denmark in 1943 he had been determined that his gold Nobel Prize medal should not fall into the hands of the Nazis. To prevent this he had dissolved it in acid, leaving behind an unobtrusive bottle of cloudy liquid on an obscure laboratory shelf. When he returned, he then precipitated the gold and recast the medal. An anecdote of exquisite poetic symbolism – it is of course too good to be true. But only just. The Nobel gold medal wasn't actually Bohr's (he had donated his to the Finnish War Relief Fund). Instead, it belonged to his old scientific sparring partner von Laue, who had sent it to Bohr for safekeeping at the start of Germany's ungolden era. (Von Laue's medal was not the only thing he had been in danger of losing. Unlike Heisenberg, von Laue had made public his opinion of the Nazis. Indeed, Nobel Prizewinners became an endangered species in

Germany during this period. In a fit of pique, Hitler had forbidden any German from accepting the Nobel Prize, because previous German winners such as Einstein, Planck, Heisenberg, Hertz, Thomas Mann etc etc were all un-acceptable to the master race.)

After the war Bohr continued his work as director of the Institute, now supported by his son Aage (who later also won the Nobel Prize, for his further development of the 'liquid drop' atomic nucleus model). After the death of Einstein in 1955 Bohr assumed the mantle of the 'greatest living scientist'. But he made good use of this absurdity – mounting a crusade for the international sharing of all knowledge about nuclear fission. In his view, this would prevent the development of even more destructive nuclear bombs. Curious logic, which failed to convince such logicians as Khruschev and Eisenhower.

Niels Bohr died in 1962 at the age of 77. That rarity, a great and a good man, he was honoured by his people who renamed his beloved institute the Bohr Institute.

SOME QUIRKS & QUARKS OF THE QUANTUM WORLD

• 'Anyone who claims that quantum theory is clear, doesn't really understand it.' Niels Bohr

• Quantum is the scientific equivalent of cubism, which sees an object from several positions at once. Quantum theory and cubism were developed simultaneously, but independently. It has been suggested that at the start of the 20th century our evolution underwent a 'jump' in the way we see the world.

• Quarks are elementary particles which combine to form protons and neutrons of the atomic nucleus. Quantum chromodynamics suggest there may be as many as 18 types of quark. These types (known as flavours) include:

up, down, strange, charmed, bottom (or beauty) and top (or truth). The term quark was coined by the American physicist Murray Gell-Mann from a word invented by James Joyce in his incomprehensible masterpiece *Finnegans Wake*, which was itself described by one critic as 'a quantum leap into the dark'.

• 'Multiculturalism is a quantum concept.' *New Scientist*

• Encyclopedia definition: 'Pluralism is the belief in the co-existence of incompatible views . . . [it] has gradually permeated all aspects of 20th century culture, society and even knowledge . . .'

• Einstein showed that the universe does not consist of matter. The ultimate particles are energy. Seen thus, all physical objects become space packed with force.

• 'Quantum is essentially science beyond sense. We can have no *picture* of ultimate reality.' Heisenberg

• 'I am sure as I can be of anything that will never be knowable that reality must be more bizarre than we shall ever be capable of conceiving.' Bryan Magee, contemporary philosopher.

• 'Profound aversion to reposing once and for all in any one total view of the world. Fascination of the opposing point of view: refusal to be deprived of the stimulus of the enigmatic' Nietzsche's prescription for the future of science in *1886*. He also said: 'The most valuable insights are *methods*.'

• 'The second law of thermodynamics decrees there can never be another Humpty Dumpty. Quantum physics renders a "theory of everything" as plausible as Father Christmas.' John Mandeville, physicist.

• A word of warning for all who seek an ultimate explanation of the world in terms of knowledge: 'Take care not to make the intellect our god; it has . . . powerful muscles but no personality.' Einstein

• A late quantum physics development (simplified version): Imagine two sub-atomic particles. At one stage this pair form a system, where the value of one balances the value of the other (such as, say: A\, B/). These particles then become separated by a huge distance (say half the universe). A is then measured and found to have value \. Thus we conclude that B must have value /. So far so simple.

But according to quantum theory, A *is without value until it is measured*. And this value also depends upon the method of measurement used. This means that when A is measured and found to be \, B, because it has once been part of the same system, must assume value /. It would have to have taken on this value *instantaneously*.

Thus if quantum theory is true, something travels faster than the speed of light. But as we know, according to Einstein's theory of relativity nothing can do that. Also, this 'ordering' (or balancing of B over a vast distance) would take place without any discernable cause. It would be beyond the realms of causality.

This phenomenon is known as the EPR Paradox.

• A further development of the EPR Paradox came with Bell's Inequality Theorem. This explained the EPR Paradox by positing an 'Un-local Reality' (ie, a real world which exists in no place). The real world that *we know* is supported by this invisible reality which remains beyond space, time or causality. According to Bell, any particles which have once been part of a system will always remain linked by this un-local reality – which is not affected by distance (however great), acts instantaneously (ie, faster than the speed of light), and forms a link that does not pass through space.

This is not the first time such a method of communication has been claimed. Thought-transference by ESP works in the same way. So also does voodoo. When you stick a pin into a model of someone who is miles away, he immediately suffers from a sharp stabbing pain in the appropriate part of his body.

However, the Bell Inequality Theorem is not only called science, but recent experiments have confirmed that the EPR paradox supports Bell's suggestion.

LIFE & TIMES OF
NIELS BOHR

1885   Niels Bohr born in Copenhagen

1900   Max Planck first formulates quantum
        theory

1903   Bohr enters Copenhagen University
        to study physics

1905   Einstein's 'annus mirabilis' – in which
        he proposes special theory of relativity
        and confirms Planck's quantum theory

1911   Bohr completes thesis on electron theory
        of metals, and goes to Cambridge to
        study under J. J. Thomson

1912   Bohr moves to Manchester to study
        with Rutherford in March. Marries
        Margarethe Norlund in August

1913   Bohr publishes theory of atomic
        structure

1914–16 Bohr at Manchester University

1916   Bohr appointed professor of
        theoretical physics at Copenhagen
        University. Einstein publishes paper
        on general theory of relativity

1921   Institute for Theoretical Physics opens
        in Copenhagen, with Bohr as director

1922   Discovery of hafnium. Bohr receives
        Nobel Prize for Physics

1924   Pauli develops his exclusion principle

1925   Heisenberg proposes matrix physics.
        Schrödinger opposes this with wave
        mechanics

1927   Heisenberg develops his uncertainty
        principle. Bohr proposes
        complementarity principle

1936   Bohr first conceives of 'liquid drop'
        model for atomic nucleus

1939   Bohr develops 'liquid drop' nucleus
        model to explain nuclear fission

1941   Heisenberg visits Bohr in Nazi-
        occupied Copenhagen, passing on
        diagram of German nuclear reactor

1943   Bohr escapes from Nazi-occupied
        Denmark

1943–45 Bohr works on Manhattan Project to

make first atomic bomb

1945   First atomic bomb dropped on
Hiroshima. Bohr returns to Denmark.
Starts campaign for international
cooperation on nuclear fission

1962   Bohr dies in Copenhagen

SUGGESTIONS FOR FURTHER READING

Ruth Moore: *Niels Bohr* (Hodder, 1967 et seq.) – a good standard biography, with a readable mixture of life and comprehensible science.

Abraham Pais: *Niels Bohr's Times* (Clarendon, 1991 et seq.) – biographical study: wide-ranging in both the life and the work. In-depth.

Niels Bohr: *Atomic Physics and Human Knowledge* (Chapman and Hall, 1958 et seq.) – an introduction to the real thing, from the horse's mouth.

Arkady Plotnitsky: *Complementarity* (Dute UP, 1994) – the new metaphysics of science, leaving common sense trailing far behind.

Also in the Big Idea *series . . .*

NEWTON & GRAVITY

Newton is one of the most influential scientists the world has ever known. Not only did he develop and formulate the theory of gravity, which gave mankind the first glimpse of the way the universe really worked, but he also discovered the concept of force, the nature of light, and changed the way we calculate. Newton's 'big ideas' were to transform the way we view the world forever.

Yet though we are all familiar with the theory of gravity (and the story of the apple falling from the tree), how many of us know how it really works? Newton's discoveries have so pervaded our everyday view that it is hard to understand how revolutionary his ideas really were. *Newton & Gravity* presents a brilliant snapshot of Newton's life and work, and gives a clear and accessible explanation of the meaning and importance of Newton's discoveries, and the way they have changed and influenced our own lives today.

HAWKING & BLACK HOLES

Stephen Hawking is perhaps one of the best-known scientists of our day: his book *A Brief History of Time* is a world-wide bestseller. His discoveries and research on black holes and cosmology have been hailed as the next leap for mankind into new worlds and a new era. The possibilities may indeed be boundless. Hawking's 'big ideas' have changed the way we view the world and the cosmos, forever.

But despite their currency in today's popular fictions – novels, films, TV series – how many of us really understand what black holes are or might mean for future generations? *Hawking & Black Holes* presents a brilliant snapshot of Hawking's life and work, and gives a clear and accessible explanation of the meaning and importance of Hawking's discoveries, and the way they may change and influence our own lives today.

EINSTEIN & RELATIVITY

$$E = mc^2$$

Few equations have entered our consciousness with the speed and impact of Einstein's cosmos-changing formula. From the moment in 1905 and 1917 he published his revolutionary papers on his Theory of Relativity, mankind's view of the world and the universe changed forever, the latest phase of the modern age was born, and our horizons shifted.

But how many of us really know what his theory really means, and what implications it has? *Einstein & Relativity* presents a brilliant snapshot of Einstein's life and work, together with their historical and scientific context, and gives a clear and accessible explanation of the meaning and importance of Einstein's Theory of Relativity, and the way it has changed and shaped our thinking in the twentieth century.

THE POWER OF READING

Visit the Random House website and get connected with information on all our books and authors

EXTRACTS from our recently published books and selected backlist titles

COMPETITIONS AND PRIZE DRAWS Win signed books, audiobooks and more

AUTHOR EVENTS Find out which of our authors are on tour and where you can meet them

LATEST NEWS on bestsellers, awards and new publications

MINISITES with exclusive special features dedicated to our authors and their titles

READING GROUPS Reading guides, special features and all the information you need for your reading group

LISTEN to extracts from the latest audiobook publications

WATCH video clips of interviews and readings with our authors

RANDOM HOUSE INFORMATION including advice for writers, job vacancies and all your general queries answered

Come home to Random House

www.rbooks.co.uk